Leica Camera

The Step-by-Step Beginner to Pro Photography Guide for Mastering Your Leica, Taking Stunning Photos, and Shooting Like a Legend

Georgette Howard

1

Table of Contents

Introduction

Welcome to Your New Leica!

There's a special kind of magic that happens when you hold a Leica in your hands. It's not just a camera, it's a masterpiece of precision, passion, and purpose. Whether this is your very first Leica or you're finally ready to unlock its full potential, congratulations! you're about to step into a world where photography becomes more than just a skill. It becomes an art form.

This guide was written for you.

- ✓ The beginner who's never picked up a "real" camera.
- ✓ The everyday explorer who's tired of complicated manuals.

9

✓ The dreamer who wants to take photos that don't just capture life, but feel like life.

✓ And yes, we'll make every step easy.

Why Leica? The Magic Behind the Brand

There's a reason Leica is the gold standard for legendary photographers around the world. From war correspondents and fashion icons to street artists and nature enthusiasts, Leica has captured some of the most iconic images in human history.

But Leica isn't just famous for its legacy. It's famous for how it makes *you feel*.

✓ Every dial, button, and lens is designed to work intuitively, smoothly, and precisely.

✓ The minimalist body invites you to slow down and see your surroundings with more clarity.

✓ The colors, the contrast, the detail—Leica cameras don't just reproduce the world; they translate it into something beautiful.

And unlike many overly technical brands, Leica believes in simplicity. That's why this guide is built the same way.

Quick Setup: Your First Photo in Minutes!

Before we dive into the deep end, let's start with something satisfying—taking your very first shot.

Here's how to get up and running, even if you've never used a manual camera before:

1. Charge the Battery

Plug it in using the supplied charger. Full charge takes about 2–3 hours.

2. Insert Your Memory Card

Use a recommended SD card (preferably Class 10 or UHS-I). Insert it gently until it clicks into place.

3. Power On

Slide the power switch or rotate the shutter button ring (depending on your model) to the ON position.

4. Set to AUTO Mode

Most Leica models have an AUTO or Program mode. Use this to let the camera choose the best settings automatically.

5. Frame Your Shot

Point your camera toward something you love. Use the viewfinder or LCD screen.

6. Gently Press the Shutter

Lightly press halfway to focus… then all the way to shoot!

That's it. You've taken your first Leica photo.

No stress. No jargon. Just the beginning of something beautiful.

This book will guide you through everything else from simple settings and composition tricks to advanced techniques and video shooting. By the end, you'll not only know how to use your Leica... you'll feel like a photographer.

So let's begin!

Let's make magic, one frame at a time.

Chapter 1

Leica Basics for Absolute Beginners

From Holding Your Camera to Taking Your First Confident Shot

Understanding Your Leica: Buttons, Dials, and Menus Simplified

Leica cameras are known for their minimalist design but beneath that simplicity lies incredible power.

If you're new to Leica (or to cameras in general), don't worry. You don't need to know what every button does right away. This guide will walk you through the essentials so you can start shooting with confidence and grow your skills over time.

Here's what you'll find on most Leica models:

Top Plate (the top of your camera)

- **Shutter Button:** This is your "go" button. Press halfway to focus, all the way to capture.

- **Mode Dial (on some models):** Choose between AUTO, Manual (M), Aperture Priority (A), etc.

- **ISO Dial:** Adjusts the camera's sensitivity to light. We'll explain how to use this later, but AUTO is perfect for now.

- **Power Switch:** Usually wrapped around the shutter button or nearby.

Back Panel

- **Viewfinder & LCD Screen:** You can compose your shots through the viewfinder or live screen. Both work great.

- **Menu Button:** Opens your camera's settings. Think of this like your "command center."

- **Playback Button:** Review your shots instantly.

- **Navigation Buttons/Dials:** Scroll through menus, adjust settings, or zoom in on photos.

- **Function Buttons (Fn):** These are customizable shortcuts. You can assign your favorite features here later.

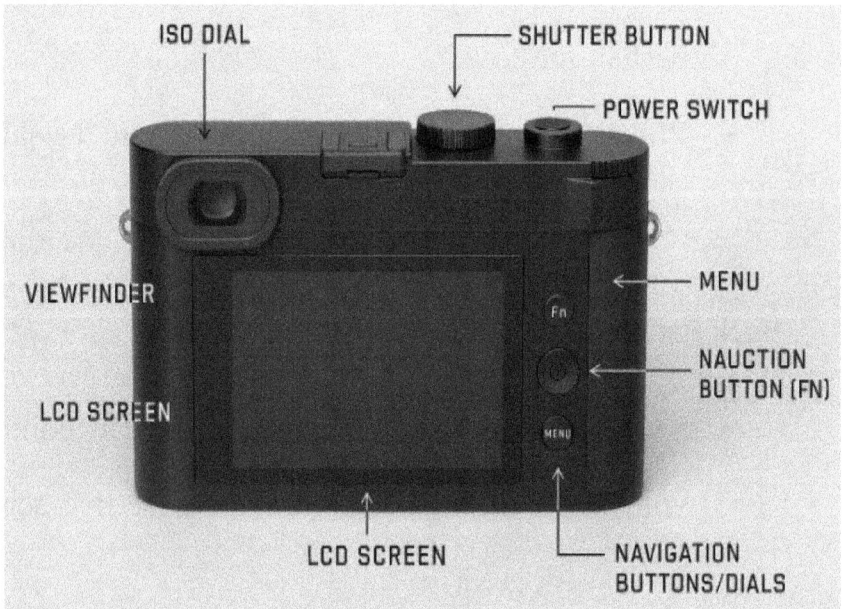

ISO DIAL | SHUTTER BUTTON | POWER SWITCH | MENU | NAUCTION BUTTON (FN) | VIEWFINDER | LCD SCREEN | LCD SCREEN | NAVIGATION BUTTONS/DIALS

Menu Navigation Tips

Don't get overwhelmed! Most Leicas have clean, tab-based menus. You'll learn which settings matter most in the coming chapters. For now, just get comfortable opening and moving around the menu.

Charging, Battery Care, and Memory Cards—Made Simple

Let's get your camera ready to shoot.

Charging the Battery

1. Use the official Leica charger or USB-C cable (depending on your model).

2. A full charge typically takes 2–3 hours.

3. Always start your day with a full battery especially if you're heading out for a shoot.

Pro Tip: Leica batteries last long, but keeping a spare is always smart.

Inserting a Memory Card

1. Power off your camera.

2. Open the memory card slot (usually on the side or bottom).

3. Gently insert the card—label facing out (depending on model).

4. If prompted, format the card through your camera's settings before use.

Recommended Cards:

Use a fast SD card—Class 10 or UHS-I/II. This ensures smooth image writing and 4K video recording if your model supports it.

First Shots: Auto Mode & Immediate Success

You're ready to shoot. Here's how to take your first photo without worrying about settings.

Step 1: Set to Auto Mode

Switch to AUTO or Program mode (P). Your Leica will handle exposure, focus, and everything technical for now.

Step 2: Frame Your Subject

Look through the viewfinder or LCD. Frame a subject that interests you—a person, a pet, your coffee mug.

Step 3: Focus

Lightly press the shutter button halfway. You'll hear a gentle click as the camera locks focus.

Step 4: Shoot!

Now press the button all the way down. That's your first Leica photo!

✅ What to Expect

- Crisp image quality

- Beautiful colors, straight out of camera

- That "Leica feel"—depth, texture, and emotion you just don't get from a smartphone

Final Thoughts for Chapter 1

Mastering your Leica doesn't start with manuals or technical jargon. It starts with holding the camera, pressing a button, and feeling the joy of capturing a moment.

You've done that now.

From here, we'll start learning how to control light, sharpen your focus, and unlock Leica's full creative potential step by step.

Chapter 2

Essential Leica Camera Settings Explained Clearly

Understand the Core Settings That Bring Your Photos to Life

Mastering Exposure: Easily Understand Aperture, Shutter Speed, and ISO

Exposure is what determines how bright or dark your photo appears and learning how to control it is the first real leap toward creative freedom.

Leica gives you full control over this process, and don't worry: we'll make it as easy as adjusting the volume on your stereo.

Exposure Triangle: Your Three Settings That Work

Together

THE EXPOSURE TRIANGLE

Aperture

Widens

Affects or narrows

Affects

Quickens or field

f/.6

ISO

100

ISO

6400

1/100 s

**Shutter
Speed**
1/100 s

Quickens or lengthens
Affects motion blur

ISO
100
6400

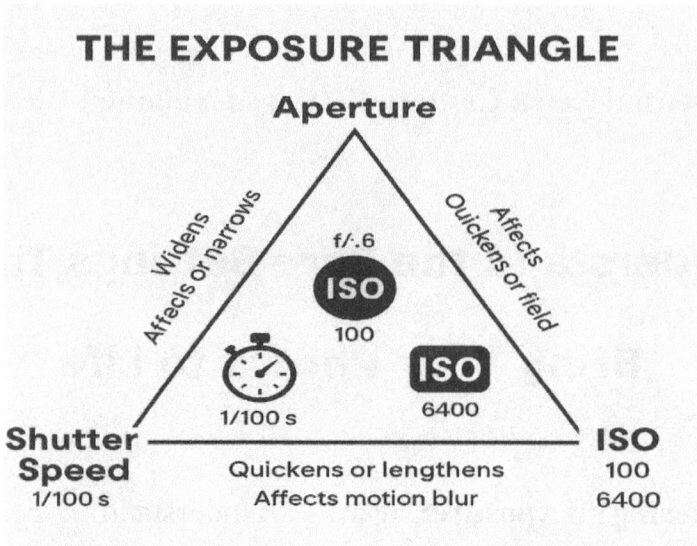

Think of **aperture**, **shutter speed**, and **ISO** as three legs of a tripod. All three must balance to create a properly exposed photo.

Aperture (f-stop): Controls the Opening of the Lens

- A **lower number** (f/1.4) = more light, and beautiful background blur (great for portraits)
- A **higher number** (f/8, f/11) = less light, but more of the image in sharp focus (great for landscapes)

22

Tip: Use low f-numbers for soft, dreamy backgrounds. Try higher numbers for sharp scenes like architecture or scenery.

Shutter Speed: How Long the Camera's Sensor is Exposed to Light

- **Fast speeds (1/500s, 1/1000s)** = Freeze action (perfect for sports or movement)
- **Slow speeds (1/30s, 1s, or more)** = Blur motion or brighten a dark scene (used creatively in night photography)

Tip: Use a tripod for shutter speeds slower than 1/60s to avoid blur from hand movement.

ISO: How Sensitive Your Camera Is to Light

- **Low ISO (100–400)** = Less grain, better quality (great for daylight)

- **High ISO (1600–6400+)** = Brighter images in low light, but with more grain or "noise"

Tip: Keep ISO low unless you're shooting in dark conditions or need a faster shutter speed.

Auto Mode Tip:

When starting out, set **Aperture Priority mode (A or Av).** You choose the aperture, and the camera picks the rest. It's a great training wheel.

Sharp and Clear Focus: Manual vs. Autofocus Simplified

A sharp image is often the difference between a snapshot and a photograph. Leica cameras excel at both manual and autofocus, let's look at both.

Autofocus (AF): Let the Camera Do the Work

24

Most Leica mirrorless models like the Q and SL series have quick and intelligent autofocus.

Simply half-press the shutter button to lock focus, then press all the way to shoot.

Use This When: You're shooting fast-moving scenes, portraits, or just want speed and ease.

Manual Focus: Total Control

On models like the M-series or when using manual lenses, you'll focus by turning the lens ring.

Focus Assist Tools:

- **Focus Peaking:** Highlights the sharpest areas in color (usually red).
- **Zoom-in Assist:** Magnifies part of your image for precise focus.

Use This When: You want precision, or when autofocus struggles (like in low light or reflective scenes).

Pro Insight: Manual focus slows you down in a good way. It helps you see more intentionally, just like a Leica is meant to be used.

Colors That Pop: White Balance Settings Made Easy

White Balance (WB) adjusts how your camera interprets colors in different types of light, so your whites look white, not yellow or blue.

Common WB Presets:

- **Auto (AWB):** Great for most situations. Leica is excellent at this.
- **Daylight:** For sunny outdoor shots.
- **Cloudy:** Warms up photos when skies are gray.
- **Tungsten:** Cools down orange indoor light.

- **Custom:** You set the color using a reference (like a white sheet of paper).

WHITE BALANCE

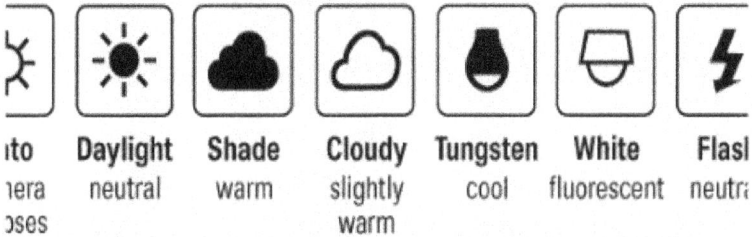

ito	Daylight	Shade	Cloudy	Tungsten	White	Flasl
iera	neutral	warm	slightly	cool	fluorescent	neutri
ɔses			warm			

COLOR TONONE

Why It Matters

Imagine taking a photo of a white plate indoors and it turns out yellow. That's white balance gone wrong.

Tip: When in doubt, use AWB. When you're editing later in Lightroom, you can still tweak it if you shoot in RAW format.

Final Thoughts for Chapter 2

- You don't need to memorize every number. What matters is this:

- Aperture controls depth and light.

- Shutter speed controls motion and sharpness.

- ISO controls brightness and grain.

- Focus makes your subject stand out.

- White Balance keeps your colors accurate.

You're no longer guessing. You're understanding. And soon, you'll be mastering.

Chapter 3

Taking Beautiful Pictures—Step by Step

From Quick Snapshots to Photographs That Tell Stories

Composition Made Simple: Easy Ways to Better Photos

The secret to a stunning photo isn't always the camera. It's how you see and how you frame the world.

That's called **composition**, and with just a few easy principles, your photos can go from "meh" to "wow."

The Rule of Thirds

Imagine your image is divided into 9 equal parts (like a tic-tac-toe board). Place your subject along the lines or at the intersections not dead center. This naturally draws the viewer's eye and adds depth.

Try This: Turn on your camera's "grid lines" in the menu. Practice placing people or objects on those points.

Leading Lines

Look for lines, roads, fences, shadows that guide the viewer's eye toward your subject. They add a sense of direction and story.

Fill the Frame

Get closer to your subject. Let it dominate the shot. This eliminates distractions and creates emotional connection.

Avoid Clutter

Too much background noise? Step left. Step closer. Simplify. Good photos are often as much about what you leave out as what you include.

Perfect Portraits Every Time: Step-by-Step Instructions

Leica cameras are incredible for portraits. They render skin tones beautifully and create that dreamy background blur that makes people pop.

Step 1: Use a Wide Aperture (f/1.4–f/2.8)

This blurs the background and sharpens your subject. It creates separation and a professional look.

Step 2: Focus on the Eyes

Always. Humans connect through eyes. If they're in focus, the whole image feels alive.

Step 3: Use Soft, Natural Light

Shade or window light is your best friend. Avoid harsh midday sun unless you're going for a dramatic effect.

Step 4: Pose Lightly

Instead of stiff "say cheese" poses, encourage movement. A small laugh. A quick turn. Let your subject be them.

Bonus Tip: Use burst mode when shooting kids or candid moments you'll always get that one magical frame.

Breathtaking Landscapes: Capture Stunning Scenery Easily

Nature deserves more than a quick snapshot. Here's how to photograph landscapes that stop people in their tracks.

Step 1: Use a Narrow Aperture (f/8–f/16)

This ensures everything from front to back is in sharp focus.

Step 2: Shoot During Golden Hour

That magical hour after sunrise or before sunset gives you soft, warm, cinematic light.

Step 3: Add Foreground Interest

Frame your mountain or beach scene with a rock, flower, or path in the foreground to add depth.

Step 4: Use a Tripod (Optional but Powerful)

Want crisp shots with zero shake, especially at slower shutter speeds? Tripods make a huge difference.

Bonus Tip: Turn on horizon level in your camera to keep your photo balanced, crooked horizons distract fast.

Fun and Engaging Street Photography Techniques

Street photography is about capturing life as it happens. Leica cameras are legendary in this genre for a reason they're quiet, compact, and discreet.

➢ **Be Invisible but Present**

Use a small lens. Avoid drawing attention. Blend into the

environment like a curious observer.

➢ Keep Moving

Walk. Pause. Watch. Life unfolds in moments—your job is to be ready.

➢ Use Zone Focusing (Manual Focus Trick)

Set your aperture around f/8 and pre-focus a few feet ahead. Everything within that range will be sharp—so you can shoot instantly without delay.

➢ Look for Human Stories

Expressions. Gestures. Contrasts between people and environment. It's not about perfect composition—it's about emotion and honesty.

Final Thoughts for Chapter 3

- You don't need to be a professional to take professional-looking photos. You just need:
- Simple composition principles

- Light you understand

- A subject that matters

- A little patience and the willingness to see

Every Leica photo you take from here on will not just document a moment... it will *tell a story*.

Chapter 4

Leica Video Basics—Shooting Like a Pro Easily

Capture Cinematic Footage Without the Complexity

Shooting video on your Leica camera isn't just about pressing "Record." It's about telling a story in motion, one that looks sharp, sounds clean, and feels effortless.

Let's break it down so you can start creating beautiful videos, whether you're filming family moments, a YouTube vlog, or a documentary-style scene.

Video Settings Simplified: Quality, Frame Rates, and More

Before you hit "record," let's walk through the basic video settings you'll want to understand. Don't worry, once you set these up once, you'll rarely need to touch them again.

Resolution (Video Size)

- **1080p (Full HD):** Great for everyday shooting. Smaller file size, easier to edit.
- **4K:** Higher detail, professional quality. Use if you want sharp, cinematic footage or plan to crop/zoom during editing.

Tip: If you're not sure, start with 1080p at 30fps.

Frame Rate (Frames Per Second – FPS)

- 24fps = Cinematic look (used in movies)
- 30fps = Smooth and natural (standard video)
- 60fps = Super smooth, great for slow motion playback

Pro Trick: Record at 60fps and slow it down in editing for beautiful slow-motion shots.

Focus Settings for Video

- Many Leica models offer Continuous Autofocus (AF-C) in video mode.

- Or switch to Manual Focus to lock in on your subject and avoid hunting focus during a clip.

Manual Focus Tip: Pre-focus on your subject and use focus peaking to lock in sharpness.

Shooting Steady Footage: Quick Tips and Tricks

Shaky video screams "amateur." The good news? Smooth footage is easy once you learn a few tricks—no gimbal required.

1. Use Two Hands & Tuck Your Elbows

Hold your Leica firmly with both hands. Keep your elbows close to your body and bend your knees slightly to act as a natural stabilizer.

2. Use Stable Surfaces or a Tripod

Rest your camera on a wall, table, or your camera bag if you don't have a tripod. Even leaning against a tree or post helps eliminate shake.

3. Slow Down Your Movements

If you're moving while filming (panning, walking), do it slowly and smoothly. Leica footage shines when handled with elegance.

4. Use Digital Stabilization (if available)

Some Leica models offer in-camera stabilization. Turn it on in your video settings to reduce jitter when shooting handheld.

Bonus: Practice slow pans moving left to right or up to down gradually. This adds motion without distraction.

Capturing Clear Audio: Simple Steps for Better Sound

A beautiful video with bad sound feels... off. Clear audio makes your footage feel professional, polished, and pleasant.

🔊 Use an External Microphone (Highly Recommended)

Leica cameras often include a mic input. Plug in a small shotgun mic or lavalier mic for dramatically improved sound.

⊘ Avoid Wind and Background Noise

Use a wind muff (also called a "dead cat") on your mic when shooting outside. Indoors, turn off fans or noisy appliances before recording.

⊺ Monitor Your Levels

Enable audio levels in your settings (if available) to ensure your voice isn't too low or peaking too high. Keep it in the green.

🎧 Use Headphones When Possible

If your camera has a headphone jack, monitor your sound as you shoot. You'll catch problems instantly instead of after the fact.

No Mic? Get close to your subject. The built-in mic will perform best within a few feet especially in quiet environments.

LEICA VIDEO BASICS

VIDEO SETTINGS SIMPLIFIED

RESOLUTION	FRAME RATE	FOCUS
4K Full HD (1080p) Good quality, smaller file	24 fps Cinematic look	Continuous AF or Manual Focus
4K 4K Higher detail, larger file	30 fps Standard 60 fps	

SHOOTING STEADY FOOTAGE

- Use two hands
- Use tripod or stable surface
- Avoid wind or noise
- Monitor audio levels
- Wear headphones

CAPTURING CLEAR AUDIO

- Use external mic
- Monitor audio levels
- Avoid wind & noise
- Wear headphones

Final Thoughts for Chapter 4

Leica makes video look luxurious and now you know how to make it feel cinematic, too. Here's what to remember:

- Choose the right **resolution and frame rate**

- Keep your shots **steady and smooth**

- Always pay attention to **your audio**

You don't need a film crew to make videos that look and sound incredible. You just need a Leica... and a little knowledge. Now you've got both.

Chapter 5

Advanced Techniques Simplified

Level Up Your Leica Photography With Low Light, Long Exposures, and Fast Action

Once you're comfortable with your Leica in everyday conditions, it's time to unlock the magic of more advanced shooting. These techniques are what separate casual snaps from stunning, gallery-worthy images, and the best part? You can do it, even as a beginner.

Taking Great Photos in Low-Light Conditions

Leica cameras are famously good in low light—but low light can still feel intimidating. Here's how to make the most of it without blur or noise ruining your shot.

44

Step 1: Use a Wide Aperture (f/1.4 – f/2.8)

A wider aperture lets in more light, making it easier to shoot in dim settings without needing flash.

Step 2: Slow Down Your Shutter (But Not Too Much)

Try 1/60s or slower. But remember: if you're handholding the camera, don't go too slow or you'll get blur. Use a tripod if you go under 1/30s.

Step 3: Increase ISO (Smartly)

Raise your ISO to 800, 1600, or even 3200 when needed. Leica's sensors handle noise well, but avoid maxing out unless absolutely necessary.

Step 4: Steady Your Shot

Use both hands, brace yourself against a wall or table, and breathe slowly before pressing the shutter.

Bonus Tip: If your camera has image stabilization, turn it on. It makes a big difference.

Easy Long Exposure Techniques for Creative Photos

Long exposures let you capture magic the eye can't see like silky waterfalls, light trails, or soft cloud motion.

Step 1: Use a Tripod (Non-Negotiable)

Any motion ruins the shot, so lock your camera down completely.

Step 2: Use Manual or Shutter Priority Mode

Set your shutter speed to 2, 5, 10 seconds—or longer. Use a remote shutter or your camera's self-timer to avoid shake when pressing the button.

Step 3: Use a Low ISO (100–200)

This reduces noise and ensures clean images during long exposure.

Step 4: Add ND Filters (Daytime Only)

Neutral Density (ND) filters darken the image, letting you shoot long exposures even in bright daylight.

Creative Ideas:

- Waterfalls and rivers = silky flow

- Busy streets = ghostly people motion

- Stars = astrophotography (use 20+ seconds)

How to Shoot Fast Action & Sports Clearly

Whether you're photographing kids playing or a fast-moving cyclist, sharp action shots are about speed, timing, and technique.

Step 1: Use a Fast Shutter (1/500s or faster)

To freeze motion, crank up the shutter speed. 1/1000s is great for very fast subjects.

Step 2: Enable Continuous Autofocus (AF-C)

Your Leica will keep tracking a moving subject great for sports, wildlife, or dynamic portraits.

Step 3: Use Burst Mode (Drive Mode: Continuous Shooting)

Capture a series of shots by holding down the shutter. Pick the best frame afterward.

Step 4: Track With Your Body

Move with your subject while shooting. It helps keep focus sharp and adds natural motion to the frame.

Pro Tip: Pre-focus on a spot where the action will happen (like a finish line) and snap as your subject enters that zone.

ADVANCED LEICA TECNIWES SIMPLIFIED

TAKING GREAT PHOTOS IN LOW-LIGHT CONDITIONS

1 USE A WIDE APERTURE (I14 –f.2.8)

2 SLOW DOWN YOUR SHUTTER (BUT NOT TOO MUCH)

3 INCREASE ISO (SMARTLY)

4 STEADY YOUR SHOT

EASY LONG EXPOSURE TECHNIQUES FOR CREATIVE PHOTOS

EASY LONG EXPOSURE TECHNIQUES FOR CREATIVE PHOTOS

1 USE A TRIPOD (NON-NEGOTIABLE)

2 USE MANUAL OR SHUTTER PRIORITY MODE

3 USE A LOW ISO (100–200)

4 ADD ND FILTERS (DAYTIME ONLY)

HOW TO SHOOT FAST ACTION & SPORTS CLEARLY

1 USE A FAST SHUTTER (1/500s OR FASTER)

2 ENABLE CONTINUOUS AUTOFOCUS (AF-C)

3 USE BURST MODE

4 TRACK WITH YOUR BODY

Final Thoughts for Chapter 5

Advanced photography isn't about complication, it's about control. Now that you understand your Leica's strengths and how to use them, you're free to get more creative, expressive, and intentional with your work.

You've gone from snapping casually to shooting with purpose.

Chapter 6

Leica Lenses—Everything You Need to Know

Choose Better, Shoot Sharper, and Care for Your Leica Glass Like a Pro

Leica lenses are more than tools, they're crafted masterpieces. Each one is designed with surgical precision to deliver stunning clarity, rich color, and breathtaking detail.

But if you're new to Leica (or photography in general), the world of lenses can feel intimidating. Let's change that. In this chapter, you'll learn which lenses to choose, which to grow into, and how to take care of them properly without the guesswork.

Choosing the Right Lens—Made

Simple

There are two main types of lenses you'll come across: prime lenses and zoom lenses. Each has its own strengths, and choosing the right one depends on what you want to shoot.

Prime Lenses (Fixed Focal Length – e.g., 35mm, 50mm)

- Known for superior sharpness and low-light performance
- Great for portraits, street photography, and everyday use
- Lightweight, simple, and elegant

Best for: Simplicity, better image quality, and practicing strong composition.

Zoom Lenses (Variable Focal Length – e.g., 24–70mm)

- Versatile and convenient—you can shoot wide and close without changing lenses
- Slightly heavier, often pricier
- Great for travel, events, and fast-moving scenes

Best for: Flexibility, travel, events, and video shooting.

How to Choose:

Ask yourself:

- Do I want **maximum image quality** and a lighter setup? → Go Prime
- Do I want **versatility** and convenience for multiple shooting situations? → Go Zoom

LEICA LENSES:
EVERYTHING YOU NEED TO KNOW

CHOOSING THE RIGHT LENS – MADE SIMPLE

PRIME

- Superior sharpness & low-light performance
- Great for portraits, street and more
- Lightweight & simple

ZOOM LENSES

- Versatile & convenient
- Great for travel, events, etc.
- Slightly heavier

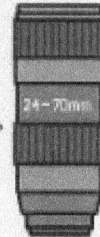

HOW TO CHOOSE:

- For maximum image quality + Go zoom
- Go for for versatility
- Go Zoom – go zuom

ESSENTIAL LENSES FOR BEGINNERS & PROS

BEGINNER-FRIENDLY LEICA LENSES

Summicron-M 50mm f/2

- Classic and versatile
- Great for travel, events, etc

PRO-LEVEL LENSES FOR THE SERIOUS SHOOTER

Noctilux-M 50mm f/0.95
- Legendary bokeh

Summilux-SL 35mm f/1.4 ASPH

Essential Lenses for Beginners & Pros

Here's a simple cheat sheet of go-to lenses—whether

you're just starting out or ready to step into the big leagues.

Beginner-Friendly Leica Lenses:

- **Leica Summicron-M 50mm f/2**

  Classic. Lightweight. Incredibly sharp. Ideal for portraits, street, and general use.

- Leica Elmarit-TL 18mm f/2.8 ASPH

  Compact and wide. Perfect for landscapes and travel with APS-C systems.

- **Leica Vario-Elmar-TL 18–56mm f/3.5–5.6**

  A great walkaround zoom for everyday situations.

Why they're great: Affordable (by Leica standards), easy to learn with, and incredibly versatile.

Pro-Level Lenses for the Serious Shooter:

- **Leica Noctilux-M 50mm f/0.95**

  The stuff of legend. Dreamy bokeh. Stunning in low light. Pricey, but jaw-dropping.

- **Leica Summilux-SL 35mm f/1.4 ASPH**

  Razor sharp and incredibly fast. A favorite for street, documentary, and video.

- **Leica APO-Summicron-SL 90mm f/2 ASPH**

  A portrait powerhouse. Gorgeous compression and color rendition.

Pro Tip: You don't need the most expensive lens to take the best photo. Know your subject, then build your lens lineup around it.

🛠Easy Lens Maintenance and Care Tips

Leica lenses are built to last a lifetime but only if you treat them right. Here's how to keep yours in top condition.

Keep Your Lens Clean (But Gently!)

- Use a blower to remove dust

- Wipe with a microfiber cloth—never your shirt!

- Use lens cleaning solution only when necessary

Always Use a Lens Cap & Hood

- Caps prevent scratches

- Hoods help with contrast and block stray light—**plus they protect the front element**

Store Lenses Properly

- Avoid high humidity (store with silica gel if possible)

- Keep in a padded camera bag or dry cabinet

- Don't leave attached to the camera for long periods when not in use

Avoid Touching the Glass

- Fingerprints leave oils that can degrade coatings over time

- Hold by the barrel or edges only

Final Thoughts for Chapter 6

Your Leica lens is the eye through which your camera sees the world. Whether it's a humble 50mm or a Noctilux masterpiece, what matters most is how you use it.

Buy smart. Shoot intentionally. Care deeply.

The better your lens—and the better you understand it— the more unforgettable your photos will become.

Chapter 7

Quick & Easy Editing and Sharing

Turn Great Shots into Gorgeous Photos and Share Them with the World

You've taken beautiful photos—now what?

The next step is where the real magic happens. Editing brings your vision to life, and sharing lets others enjoy your work. In this chapter, you'll learn how to quickly transfer, enhance, and share your Leica shots without frustration even if you're not "tech-savvy."

Transferring Your Photos to Your Computer or Phone

First, let's get your images off your camera and onto a

device where you can view, organize, and edit them.

Option 1: Transfer to Smartphone (Wirelessly with Leica App)

Most modern Leica cameras (like the Q2, Q3, SL2) support wireless transfer.

Steps:

1. Download the Leica FOTOS app (available on iOS & Android).

2. Turn on Wi-Fi on your camera and connect it to the app via Bluetooth/Wi-Fi.

3. Select the images you want and import them directly to your phone.

Tip: This is perfect for quick edits and social media sharing.

Option 2: Transfer to Computer (USB or Card Reader)

Steps:

1. Use the USB-C cable to connect your camera to your computer.

2. Or remove the SD card and insert it into your computer's card reader.

3. Open the folder and copy the images to your desired location.

RAW or JPG?

- **RAW files (DNG)** = best for editing (larger, higher quality)
- **JPG files** = smaller, ready to use (great for quick sharing)

Quick Editing for Stunning Results

Editing doesn't have to be complicated or time-consuming. With just a few tweaks, you can make your images pop while staying true to your original vision.

Best Beginner-Friendly Editing Apps & Tools

- **Smartphone:**
  - Snapseed (iOS/Android) – Free, powerful, easy to use
  - Lightroom Mobile – Professional-grade edits with presets

- **Desktop:**
  - *Lightroom Classic* (Mac/Windows) – Industry standard for photographers
  - *Luminar Neo* – AI-powered, beginner-friendly alternative
  - *Photos App* – Built-in editors on macOS and Windows work surprisingly well!

Simple Steps to Edit Like a Pro (In Under 5 Minutes)

1. Crop & Straighten – Improve composition instantly.

2. Adjust Exposure – Brighten or darken the photo.

3. Boost Contrast & Clarity – Add depth and sharpness.

4. Enhance Colors (Vibrance/Saturation) – Make it pop, not fake.

5. Apply a Preset or Filter – For consistent style (but don't overdo it).

6. Sharpen Slightly – Just enough to make details stand out.

Tip: Less is more. Great editing is about enhancement, not transformation.

Sharing Photos on Social Media Effortlessly

You've taken and edited a beautiful image. Now it's time to share it.

Best Platforms to Share Your Leica Photos

- **Instagram** – Perfect for showcasing work to a wide audience

- **Facebook** – Great for sharing with friends, family, and communities

- **Flickr or 500px** – For serious photographers & curated portfolios

- **WhatsApp/Email** – Simple, direct sharing to individuals

Tips for Effective Social Sharing

- Always **resize** large images (especially RAW) before uploading

- Add a **caption or story** behind the image—it builds connection

- Use **hashtags** like #LeicaCamera, #LeicaPhotography, #LeicaLove to increase reach

- Stay consistent with your **editing style** to build a visual brand

Bonus: Create a Backup Plan

- Always back up your photos! Use:

- An external hard drive

- Google Photos, iCloud, Dropbox

- Adobe Creative Cloud storage

- Your photos are valuable, keep them safe.

Quick & Easy Editing and Sharing

Transferring Your Photos to Your Computer or Phone

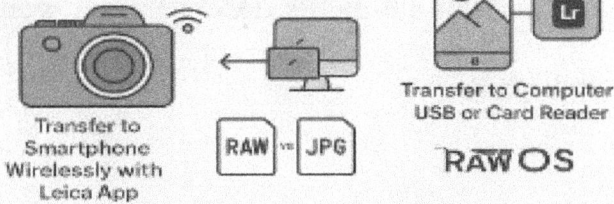

Transfer to Smartphone Wirelessly with Leica App

RAW vs JPG

Transfer to Computer USB or Card Reader

RAW OS

Quick Editing for Stunning Results

Beginner-Friendly Editing Apps & Tools

Simple to Steps to Edit like a Pro

Adjust Exposure   Enhance Colors   Apply a Preset

Best Platforms to Share Your Leica Photos

Sharpen   Slightly

Sharing Photos on Social Media Effortlessly

Final Thoughts for Chapter 7

Photography isn't finished when you press the shutter. It continues through thoughtful editing and intentional sharing.

Now you know how to:

- Transfer files effortlessly

- Make edits that elevate your images

- Share with confidence and creativity

- Your Leica moments deserve to be seen and now, they will be.

Chapter 8

Accessories—What You Really Need

Only the Essentials to Protect, Support, and Enhance Your Leica Experience

In the world of photography, it's easy to fall into the trap of buying every shiny new gadget. But the truth is, most Leica photographers only need a few thoughtfully chosen accessories to unlock the full potential of their camera, without the clutter or cost.

This chapter keeps it simple and effective: what to buy, why it matters, and how it makes your Leica life easier, smoother, and more enjoyable.

Best Tripods, Bags, and Cases—Made Simple

Accessories aren't just about convenience, they protect your investment and improve the quality of your shooting experience.

Tripods: For Steady, Sharp Shots

A good tripod eliminates camera shake, enables long exposure, and makes shooting more intentional.

What to Look For:

- Lightweight but sturdy (carbon fiber is ideal for travel)
- Quick-release plate for easy mounting
- Adjustable height with smooth leg locks
- Ball head for flexible movement

Top Picks:

- *Peak Design Travel Tripod* – Compact, premium quality
- *Manfrotto Befree* – Great balance of price and performance

Tip: Even a mini desktop tripod is great for vlogging or

stable low-angle shots.

Camera Bags & Slings: Carry with Confidence

You need a bag that protects your Leica without slowing you down.

What to Look For:

- Padded compartments
- Weather-resistant fabric
- Compact size (Leicas are small, your bag should be too)
- Quick-access design

Top Picks:

- *Billingham Hadley Small* – A classic for Leica lovers
- *Peak Design Everyday Sling* – Versatile, minimalist, secure

Protective Cases & Covers

Protect your camera body from scratches, dings, and weather.

Recommended:

- Leather half cases – Stylish + functional
- Silicone covers – Grippy and shock-absorbing
- Screen protectors – Prevent scratches on the LCD

Using Flashes Easily: Quick Guide for Better Photos

Leica's natural light rendering is beautiful but in low light or for portraits, a well-used flash can dramatically improve your results.

When to Use a Flash:

- Indoor events

- Night portraits

- Fill light during daylight (to remove harsh shadows)

Flash Options for Leica:

- Leica SF 40 or SF 60: Compact, reliable, compatible with TTL (auto flash control)

- Godox V1 or TT350: Affordable third-party options with great performance

- External Trigger + Off-Camera Flash: For advanced setups

Quick Flash Tips:

- Bounce the flash off the ceiling for soft, natural light

- Use a diffuser (or a small piece of white fabric) to soften harsh light

- Lower flash power to avoid overexposed faces

Bonus Tip: Use flash manually at low power for creative light control.

Accessories—What You Really Need

Best Tripods, Bags, and Cases—Made Simple

Tripod
- ✓ Lightweight
- ✓ Easy to adjust
- ✓ Ball head

Camera Bags Slings
- ✓ Padded
- ✓ Weather-resistant

Protective Cases
- ✓ Leather half-cases
- ✓ Silicone covers
- ✓ Screen protectors

Using Flashes Easily: Quick Guide for Better Photos

Indoor events

- Indoor events
- Night portraits
- Fill light

Third-party flashes

Recommended Accessories to Enhance Your Experience

Lens Pen & Air Blower

High-Speed SD Cards

Leica FOTOS App

Remote Shutter / Interval Timer

Recommended Accessories to Enhance Your Experience

You don't need a drawer full of gear—but here are a few thoughtful additions that make a big difference.

Lens Pen & Air Blower

- For keeping your lens and sensor dust-free

- Essential for image clarity

Extra Batteries

- Leica battery life is solid, but always carry a spare

- Especially useful for travel or long shoots

High-Speed SD Cards

- At least UHS-I, Class 10 or faster

- 64GB or 128GB for everyday shooting

- Brands: SanDisk, Lexar, Sony

Leica FOTOS App

- Wirelessly control your camera, transfer images, and geotag photos from your phone

- Great for remote shooting, family portraits, or discreet street shots

Remote Shutter Release or Interval Timer

- For long exposures, self-portraits, or timelapse photography
- Removes the risk of touching the camera and causing blur

Final Thoughts for Chapter 8

With Leica, less is more. You don't need dozens of accessories, just a few purposeful, high-quality tools that match the way you shoot.

Protect what you love. Support your creativity. And carry only what matters.

Chapter 9

Caring for Your Leica—Easy Tips for Long Life

Keep Your Camera in Peak Condition with Simple, Smart Habits

Your Leica is more than a camera, it's an investment, a companion, and a legacy tool that can last decades when cared for properly.

Thankfully, you don't need to be a technician to maintain your camera. This chapter gives you easy, no-stress practices that will extend your Leica's life, improve its performance, and ensure it's always ready to capture the moment when it matters most.

Cleaning & Basic Maintenance Made

Easy

Dirt, dust, smudges, and moisture can slowly degrade your camera's performance. With a simple routine, you can avoid costly issues down the road.

What You'll Need

- Air blower

- Lens cleaning pen (brush + microfiber tip)

- Microfiber cloth

- Lens cleaning solution (optional)

- Sensor swab kit (only if absolutely necessary)

How to Clean Your Camera Safely

1. Exterior:

Use a dry microfiber cloth to gently wipe the camera body. Avoid water or harsh chemicals.

2. Lens:

- Use the air blower first to remove loose dust.

- Gently brush with a lens pen.

- If needed, use a drop of lens solution on a cloth never directly on the lens.

3. Viewfinder & LCD Screen:

- Wipe with a dry microfiber cloth.

- Use a screen protector to prevent scratches.

4. Sensor:

- If you see spots on your images, your sensor may have dust.

- Use a sensor blower first. Only use a sensor swab if you're comfortable or have it cleaned by a pro.

Pro Tip: Clean in a dust-free, indoor environment. Never clean when the camera is powered on.

Protecting Your Leica—Best Practices

Good habits go a long way in preventing damage before it happens.

Avoid Moisture and Humidity

- Never leave your Leica in a damp bag or car.
- Use silica gel packs in your bag or case to absorb moisture.
- Store in a dry cabinet or sealed container during long storage.

Avoid Extreme Temperatures

- Don't expose your Leica to prolonged direct sun, freezing cold, or sudden temperature shifts.
- Let your gear acclimate before using it after going from cold to warm (e.g., winter shooting into a warm room).

Use a Case or Camera Bag

- Always store your camera in a padded bag.

- A leather half-case can protect against scratches during daily use.

Battery & Charging Habits

- Remove battery if storing the camera for more than a few weeks.

- Avoid completely draining the battery frequently.

- Use the official Leica charger when possible.

🔧 Troubleshooting Common Issues Quickly

Most Leica issues are simple to fix on your own. Here's a quick guide to stay in control.

Blurry Photos?

- Check if your lens is clean.

- Increase shutter speed or stabilize your camera.

- Make sure autofocus is active and correctly set.

Dust on Images?

- Try blowing the sensor with a blower.

- If it persists, it's time for a sensor cleaning or service.

Battery Not Charging?

- Try a different charger and cable.

- Clean the battery contacts gently with a dry cloth.

- If still unresponsive, replace the battery—it may be at end of life.

Camera Freezing or Lagging?

- Turn off and remove battery for 30 seconds.

- Reformat the SD card (after backing up your files).

- Update your camera's firmware via Leica's website.

Tip: Keep your firmware updated to avoid bugs and gain new features.

CARING FOR YOUR LEICA—EASY TIPS FOR LONG LIFE

CLEANING & BASIC MAINTENANCE MADE EASY

Air blower   Microfiber cloth   Lens cleaning solution

What You'll Need

1. Exterior: Wipe with a dry microfiber cloth
2. Lens: Remove dust with air blower & blush; Wipe with pen or cloth
3. Viewfinder & LCD: Clean with microfiber cloth only
4. Sensor: Use air blower; Sensor swab kit if neded

TROUBLESHOOTING COMMON ISSUES QUICKLY

Battery   Camera

PROTECTING YOUR LEICA— BEST PRACTICES

Avoid Moisture and Humidity   Avoid extreme temperatures

Use a Case or Camera Bag   Battery & Charging Habits

TROUBLESHOOTING COMMON ISSUES QUICKLY

Blurry Photos?   Dust on Images?

Battery Not Charging?   Camera Freezing or Lagging?

Final Thoughts for Chapter 9

Caring for your Leica isn't about being a camera geek, it's about respect. Respect for craftsmanship, for legacy, and for the moments you're entrusted to capture.

By staying clean, mindful, and prepared, you're not just maintaining a tool, you're nurturing a creative partner that will serve you faithfully for years to come.

Chapter 10

Professional Secrets Simplified

Unlock Expert-Level Results Without the Overwhelm

By now, you've covered the foundations—how to shoot, edit, care for your gear, and share your Leica images with pride.

Now it's time to go one step further.

This chapter reveals the little things pros do that make a big difference. You'll also discover smart shortcuts, hidden camera features, and quick ways to fix the most common mistakes so you can shoot like a pro without years of trial and error.

Easy Pro Tips for Stunning Leica

Photography

These are the tried-and-true techniques professional Leica users swear by, yet they're simple enough for anyone to start applying today.

Master Light Before Gear

- Pro secret: Light matters more than lenses.
- Shoot during **golden hour** (1 hour after sunrise or before sunset) for rich, warm tones.
- Look for **natural reflectors** (walls, sand, pavement) to bounce light gently onto your subject.

Shoot with Intention

- Don't just "take" a photo—make it.
- Ask: What am I trying to say with this frame? What emotion or focus do I want?

- Adjust your composition, background, and settings accordingly.

Use Manual Settings to Shape Your Image

- Learn when to override AUTO. Use manual focus or exposure to create drama, contrast, or mood.

- Pro photographers use full control to sculpt their images—not just expose them.

Slow Down

- Leica cameras are made for mindfulness.

- Instead of 100 quick shots, try 10 thoughtful ones.

- You'll get closer to your subject and your results will show it.

Hidden Features Every User Should Know

Leica cameras often hide genius features in their

minimalist menus. Here are a few that pros love and many beginners overlook.

Focus Peaking

- Highlights in-focus edges (usually in red, blue, or white).
- Great for precise manual focus—especially in low light or macro work.

Magnified Focus Assist

- Zooms into your focus area when manually adjusting focus.
- Ideal for portraits or detail shots where sharpness matters.

Exposure Simulation (Live View)

- Shows how your exposure settings will look before you take the photo.
- Helps you learn fast and avoid over/underexposing shots.

Custom Profiles (User Settings U1/U2/U3, etc.)

- Save your favorite configurations for quick recall.

- Example: U1 = Portraits, U2 = Street, U3 = Video

- Set once and save time forever.

Leica FOTOS App Remote Shooting

- Control your camera wirelessly via smartphone.

- Great for self-portraits, discreet street photography, or capturing wildlife.

Quickly Solve Common Mistakes—A Simple Guide

Even pros mess up. The difference? They know how to recover fast. Here's a cheat sheet for common Leica problems and quick fixes.

Your Shots Look Dull or Flat?

- Boost contrast slightly in-camera or in post

- Use better light harsh sun or dim shadows make images feel lifeless

- Add depth by using a wider aperture (f/2.0–f/4.0)

Photos Are Blurry?

- Raise your shutter speed (1/250s or faster)

- Use image stabilization if available

- Check your focus point—recompose if needed

Subject Is Out of Focus?

- Use single-point autofocus or manual focus

- Focus on the eyes for portraits

- Avoid focus-and-recompose at very shallow depth of field (f/1.4–f/2.0)

Images Are Too Dark or Too Bright?

- Use exposure compensation (+/- button)

- Watch the histogram in live view

- Enable highlight warnings ("zebra stripes") in settings

You Feel Stuck or Uninspired?

- Limit yourself to one lens for a week

- Try black and white shooting mode

- Look for patterns, shadows, symmetry—creative exercises spark growth

Final Thoughts: Your Leica Journey Has Just Begun

The beauty of Leica is that it doesn't rush you—it invites you to see more deeply, shoot more intentionally, and create with purpose.

You've learned the tools. You've mastered the fundamentals. And now, you're equipped with the mindset and skillset to create images that matter.

So go out.

Explore.

Observe.

And trust your Leica to help you feel your way into the frame.

This isn't the end.

It's the beginning of a more meaningful photographic journey.

Acknowledgement

To every aspiring photographer and filmmaker who dares to pick up a camera and tell a story, this book is for you. Special thanks to my family and friends for their encouragement, and to the creative community whose passion inspires me daily. Your support made this guide possible.

www.ingramcontent.com/pod-product-compliance
Lightning Source LLC
Chambersburg PA
CBHW031907200326
41597CB00012B/545